The Exact QED Calculation of the Fine Structure Constant Implies ALL 4D Universes have the Same Physics/Life Prospects

Exact QED Calculation of Fine Structure Constant α
Approximate Formula in Terms of π and e
Fine Structure Constant is Independent of Time and Space
Universe Vacuum Polarization Generates Expansion
Trivializes Anthropic Principle
Physics and Chemistry of 4D Universes
Life Can Exist in All 4D Universes

Stephen Blaha Ph. D.
Blaha Research

Pingree-Hill Publishing
MMXIX

Cover: The cover table depicts the fine structure constants of QED and the Other Known Interactions. It also shows a plot of the QED eigenvalue function displaying the point where α appears.

Rev. 00/00/01   September 15, 2019

To Margaret

Some Other Books by Stephen Blaha

All the Megaverse! Starships Exploring the Endless Universes of the Cosmos using the Baryonic Force (Blaha Research, Auburn, NH, 2014)

SuperCivilizations: Civilizations as Superorganisms (McMann-Fisher Publishing, Auburn, NH, 2010)

All the Universe! Faster Than Light Tachyon Quark Starships & Particle Accelerators with the LHC as a Prototype Starship Drive Scientific Edition (Pingree-Hill Publishing, Auburn, NH, 2011).

Cosmos Creation: The Unified SuperStandard Model, Volume 2, SECOND EDITION (Pingree Hill Publishing, Auburn, NH, 2018).

Immortal Eye: God Theory: Second Edition (Pingree Hill Publishing, Auburn, NH, 2018).

Unification of God Theory and Unified SuperStandard Model THIRD EDITION (Pingree Hill Publishing, Auburn, NH, 2018).

Calculation of: QED α = 1/137, and Other Coupling Constants of the Unified SuperStandard Theory (Pingree Hill Publishing, Auburn, NH, 2019).

Coupling Constants of the Unified SuperStandard Theory SECOND EDITION (Pingree Hill Publishing, Auburn, NH, 2019).

Available on Amazon.com, bn.com Amazon.co.uk and other international web sites as well as at better bookstores (through Ingram Distributors).

CONTENTS

FIGURES and TABLES

INTRODUCTION

Recent studies of vacuum polarization have led the author to consider their profound implications for our universe and other universes in the Megaverse. *Much more detail on the exact calculation of the Fine Structure Constant, and also on universe vacuum polarization is presented.*

In a sense this book is a supplement to *Coupling Constants of the Unified SuperStandard Theory SECOND EDITION* and *Quantum Big Bang-Quantum Vacuum Universes (Particles)*. It shows that the combined studies have a common basis in vacuum polarization and consequently we can describe the nature and growth of our universe and other four-dimensional universes in detail.

We will see the expansion/contraction of our universe (and other 4D universes) is governed by Dark Energy whose origin is in a vector field that generates vacuum polarization through a standard field theory mechanism. The mechanism is based on an eigenvalue function that has an unanticipated universality: it determines the QED fine structure constant, the coupling constants of the Weak SU(2) and Strong SU(3) gauge theories to good approximation, and the coupling constant of the QED-like vector field that causes universe expansion through Dark Energy vacuum polarization.

The result is that our universe is very similar to other universes with the same Physics and Chemistry, and thus the same prospects for life.

Chapter 1 describes our Fine Structure eigenvalue condition that we found in 1973. Chapter 2 describes our exact QED Fine Structure Constant calculation (to the known 13 place accuracy) with $F_2 \cong 0$ in much greater detail than our previous books. It shows our "approximate" eigenvalue condition appears to be remarkably accurate. Chapter 3 describes the extension of our eigenvalue function to the interactions of the Standard Model of elementary particles. Chapter 4 describes the extension of our eigenvalue function to the evolution of universes. It shows that the universe scale factor has the same form and numerics as the fourier transform of the scalar boson (universe particle) vacuum polarization. Chapter 5 discusses the implications of the results of previous chapters. We see that 4D universes are like particles. Further, the Standard Model and universe scale factor successes show other 4D universes (and perhaps those of higher dimension) have the same Physics (and Chemistry) and thus the same prospects for life. All 4D universes are Anthropic.

1. Universal Coupling Constant Eigenvalue Condition

In a series of remarkable papers Johnson, Baker and Willey[1] developed a finite theory of massless QED (called JBW) without divergences if a certain function $F_1(\alpha)$ of the fine structure constant α called the eigenvalue function were zero. (A zero would imply Z_3, the divergent vacuum polarization constant of the electron, was zero.)

Adler[2] refined the discussion by pointing out that a zero of the eigenvalue function would be an essential singularity with:

$$F_1(\alpha) = 0 \qquad\qquad (1.1)$$
$$d^n F_1(\alpha)/d\alpha^n = 0$$

The calculation of the eigenvalue function was reduced by JBW to the sum of all single loop vacuum polarization diagrams of the general form of Fig. 1.1.

In 1973 the author[3] calculated $F_1(\alpha)$ approximately to all orders in α. A search for an essential singularity proved fruitless. Recently the author noticed that the vacuum polarization of the electron is manifest in experiment with the effective value of α increasing at higher energies. Thus Z_3 is not zero and has a divergent piece.

On this basis the author proposed, in a series of books in 2019, that the *appropriate* eigenvalue condition was

$$F_2(\alpha) = 0$$

where

$$F_2(\alpha) = F_1(\alpha) - [2/3 + \alpha/(2\pi) - (1/4)[\alpha/(2\pi)]^2] \qquad (1.2)$$

---

[1] Summarized in some detail in K. Johnson and M. Baker, Phys. Rev. **D8**, 1110 (1973). Also in Blaha (2019b) and (2019c).
[2] S. Adler, Phys. Rev. **D5**, 3021 (1972).
[3] S. Blaha, Phys. Rev. **D9**, 2246 (1974).

The additional terms are those appearing in the exact low order calculation of $F_1(x)$:

$$F_{1\text{ low order}}(\alpha) = 2/3 + \alpha /(2\pi) - (1/4)[\alpha/(2\pi)]^2 \qquad (1.3)$$

In terms of F_2 the renormalization constant Z_3 is

$$Z_3 = 1 + F_1(\alpha)\ln(p/\Lambda) = 1 + F_2(\alpha) + \text{divergent terms} = 1 + \text{divergent terms} \qquad (1.4)$$

The original goal of the JBW Model was to solve massless QED in a manner that made all renormalization constants either 1 or at least finite.

We modified this goal. We shall see that we can obtain a physically better eigenvalue function F_2 that has a zero at the known fine structure constant α. Until now we have not specified the value α that appears in the preceding equations. We now define α as a partially renormalized quantity that is related to the bare fine structure constant α_0 by

$$\alpha = \alpha_0[2/3 + \alpha_0/(2\pi) - (1/4)[\alpha_0/(2\pi)]^2] \qquad (1.5)$$

We will show in chapter 2 that the evaluation of the F_2 eigenvalue function gives the known approximate[4] physical value[5] of the fine structure constant:

$$\alpha = 0.007297352\ 5\ \ 693\ (11) \qquad (1.6)$$

The renormalized expressions appearing below are not fully finite. However the intermediate renormalized finite α is physically sensible—more so than the completely finite renormalization constants goal of the JBW Model.

The bare charge constant α_0 is known to approach ∞ at very short distances. The simplest examples of this phenomenon are the physical Coulomb scattering amplitudes

---

[4] The constant α is an irrational number.
[5] 2018 CODATA: P. J. Mohr *et al* CODATA group (2019)

and the first order change in hydrogen-like atomic energy levels.[6] Thus our modified JBW Model with a partial renormalization conforms to physical reality:

$$Z_3 = 1 + \{\alpha F_2(\alpha) + \alpha[2/3 + \alpha/(2\pi) - (1/4)[\alpha/(2\pi)]^2]\}\ln(p/\Lambda)$$
$$= 1 + \alpha\{2/3 + \alpha/(2\pi) - (1/4)[\alpha/(2\pi)]^2\}\ln(p/\Lambda) \qquad (1.7)$$

at α = the physical fine structure constant where $F_2(\alpha) = 0$.

Our approximate 1973 solution, which summed one loop pieces of the vacuum polarization yielded the algebraic equations, is:[7]

$$A_1 = (g + 1)(1 - 2g^2)/[(g + 2)(g - 1)] \qquad (1.8)$$

$$A_2 = [8g^2(2g + 1) - (2g^3 + 2g^2 + g - 2)(g^2 + 2g + 2)]/[2(g^2 - 1)(g^2 - 4)]$$

$$A_3 = -2(1 + 3g + 6g^2 + 2g^3)/[g(g + 1)]$$

$$A_4 = -(g + 2)(1 + 5g + 6g^2 + 2g^3)/[g(g^2 - 1)] - 1/(g + 1)$$

$$\psi = [gA_3 - (4 + 2g)A_1]/[(4 + 2g)A_2 - g A_4]$$

$$(\alpha/2\pi) = [gA_4 - (4 + 2g)A_2]/(A_4A_1 - A_2A_3)$$

$$F_1(g) = (2/3)(1 - 3g^2/2 - g^3) - (\alpha/4\pi)[(2 + 4g + 4g^2)(g - 2) + \alpha\psi g^3]/[(g^2 - 1)(g - 2) + \alpha(2 + 4g + 4g^2)(g - 2) + \alpha\psi g^3]$$

expressed as a function[8] of g (the power of the divergent factor p/Λ) with ψ specifying the gauge, and with the renormalization definitions

---

[66] See E. A. Ueling, Phys. Rev. **48**, 55 (1935) and R. Serber, Phys. Rev. **48**, 49 (1935).

[7] Blaha *op. cit.*

[8] The solution for the eigenvalue function is clearly best expressed in terms of the g factor in the exponents of the divergent renormalization factors. We use $F_1(g)$ and $F_1(\alpha(g))$ interchangeably.

$$\Gamma_\mu(p) = f(\gamma_\mu + 2g\gamma \cdot pp_\mu/p^2)(p/\Lambda)^{2g} \tag{1.9}$$
$$S_F = [f\gamma \cdot p(p/\Lambda)^{2g}]^{-1} \tag{1.10}$$
$$\Gamma_{\mu\alpha}(p) = (f_3/p^2)(\gamma \cdot p\gamma_\mu\gamma_\alpha - \gamma_\alpha\gamma_\mu\gamma \cdot p)(p/\Lambda)^{2g} \tag{1.11}$$

and

$$F_1 = (2/3)(1 - 3g^2/2 - g^3) - f_3/f \tag{1.12}$$

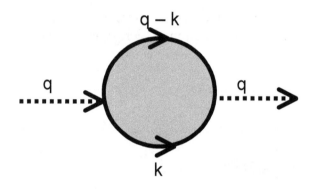

Figure 1.1 One loop vacuum polarization Feynman diagrams with internal free photon propagators.

We thus have an expression for the eigenvalue function F_2 within the framework of massless QED. We shall see that the eigenvalue function generalizes to all Standard Model gauge interactions. We shall also see that it applies to the expansion and contractions of universes upon the introduction of a Dark Energy gauge interaction for universes.

In Blaha (2019b) we generalized eqs. 1.8 to a "universal" eigenvalue function F_2 to include the Weak interaction and the Strong interaction coupling constants by inserting an interaction specific factor in the α_G equation:

$$(\alpha_G/2\pi) = c_G^{-1}[gA_4 - (4 + 2g)A_2]/(A_4A_1 - A_2A_3) \tag{1.13}$$

For a non-abelian group we set
$$c_G^{-1} = [(11/3)C_{ad} - 2C_f/3]/(16\pi)^3 \tag{1.14}$$

where C_{ad} is the dimension of the fundamental representation of the group and C_f is the number of fermions (fermion flavor) of the interaction.

We also extend it to the universe Dark Energy interaction in Blaha (2019c).

In chapter 2 we will see that we obtain the exact QED Fine Structure Constant to the known 13 place accuracy with $F_2 \cong 0$. Thus our "approximate" eigenvalue condition $F_2(\alpha) = 0$ appears to be remarkably accurate. It generalizes directly in chapters 3 and 4 to the Standard Model interactions and the universe scale factor.

Blaha (2019b) describes the universal eigenvalue function F_2 in detail including numerous plots. It also proposes a generalization of F_2 to an "exact" tangent form in chapter 9.

2. QED α Eigenvalue Calculation

We have examined the values of the quantities in eq. 1.8 looking for an essential singularity (eq. 1.2) or its approximation. Fig. 2.1 below plots $F_2(\alpha)$ as a function of g. It displays a "flat region." While essential singularities usually are thought to imply a transcendental function such as $\exp(1/\alpha)$, a constant function with value zero fulfills the essential singularity conditions in eq. 1.1. Therefore we take the "flat region" to indicate an essential singularity.

Fig. 2.2 shows a "close up" of the flat region[9] where F_2 is approximately zero. Upon close numeric analysis we find the results in Tables 2.1 and 2.2.

g =	-0.0005805369 0000	-0.0005805369 1948	-0.0005805369 5000
α =	0.007297352	*0.0072973525693*	0.007297353
$F_2 \times 10^{10}$ =	3.26316 06817671	3.26316 025452474	3.26316 134861337

Table 2.1. Values of g, α and $F_2(\alpha) \times 10^{10}$. F_2 is very close to zero for the displayed range of values and throughout the flat region. F_2 has a local **minimum** at precisely the known value of α = 0.0072973525693 (11).

g =	-0.00058053700	-0.00058053705	-0.00058053710
α =	0.007297354	0.007297354	0.007297355
$F_2 \times 10^{10}$ =	3.26316 299072544	3.26316 29663526	3.26316 408259273

Table 2.2. Other neighboring values of g, α and $F_2(\alpha) \times 10^{10}$ in the flat region *away* from g = 0 (where our approximate F_2 is exactly zero.) F_2 is very close to zero for the displayed range of values and throughout the flat region.

---

[9] These figures appeared in Blaha (2019a) and (2019b).

Thus we have a very good approximation $F_2(\alpha) \cong 0$ at the experimentally known value that is exact to 13 places with a minimum in $F_2(\alpha)$ as anticipated.

F_2 is nearly zero, as are its derivatives, at the physical Fine Structure Constant. It closely approximates a trivial essential singularity of constant value zero in a neighborhood of the singularity.

Note $F_2(\alpha = 0) = 0$ as well. This zero can be viewed as a type of singularity. If QED could transition from positive α to negative α then it would lead to a catastrophe since like charges would then attract.[10,11]

It is extremely important to note the calculation is strictly QED. Thus α is space and time independent, and not Anthropic.

---

[10] Freeman Dyson has speculated on this possibility.
[11] $F_2(\alpha)$ may have more than one zero. One of the zeroes is at the value of the Fine Structure Constant as we show.

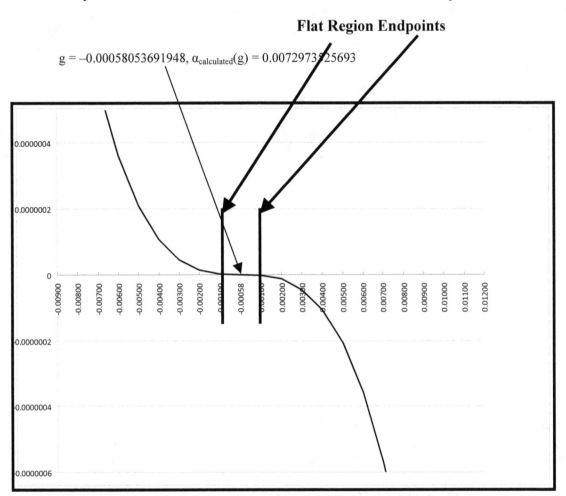

Figure 2.1. Close up plot of our eigenvalue function $F_2(g)$ (vertical axis) vs. g.

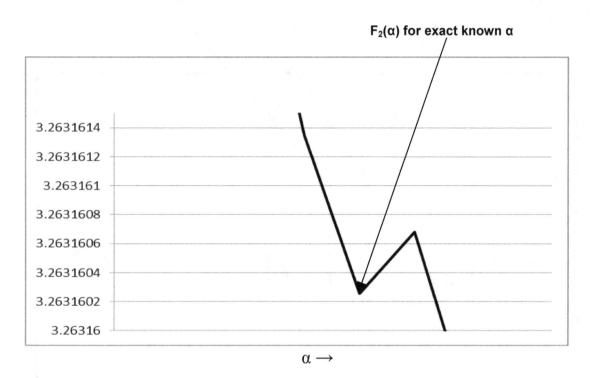

Figure 2.2. Detailed closeup plot of $F_2(\alpha) \times 10^{10}$ data in Tables 2.1 and 2.2. The local minimum of $F_2(\alpha) \times 10^{10}$ at g = -0.00058053691948 corresponds to the exact known value of α = 0.0072973525693.

2.1 Approximate Formula for α in Terms of π and e

Our success in exactly determining α is very encouraging and suggests our approximate $F_2(\alpha)$ calculation in QED may have captured the essence of the full eigenvalue function. However it is of interest to consider α from a different perspective—approximate formulas based on fundamental mathematical constant such as π and e. Speculations of simple formulas for α have been considered for over 100 years.

We shall develop two expressions for α based on

$$e = 2.718281828$$
$$\pi = 3.141592654$$

A remarkably simple expression, yet quite accurate, is

$$\alpha = 1/(16\pi e) \tag{2.1}$$
$$= 0.007318729$$

It is accurate to three places.

A better formula, to which the above is an approximation, is

$$\alpha = 1/[16\pi e(1 + \alpha/e)] \tag{2.2}$$

Solving for α we obtain

$$\alpha = (-e + (e^2 + 1/4\pi)^{1/2})/2 \tag{2.3}$$
$$= 0.007299129$$

giving five place accuracy compared to the known value:

.

$$\alpha = 0.0072973525693 \,(11)$$

3. Standard Model Coupling Constant Eigenvalues

The remarkable calculation of the QED Fine Structure Constant as evidenced by Table 2.1 leads us to consider the determination of the coupling constants of the other Standard Model interactions; weak SU(2) and Strong SU(3).

In Blaha (2019b) we generalized the F_2 function (and related functions in eq. 1.8) to the cases of the Weak interaction and the Strong interaction coupling constants by inserting a group theoretic factor in the equation for α_G only:

$$(\alpha_G/2\pi) = c_G^{-1}[gA_4 - (4 + 2g)A_2]/(A_4A_1 - A_2A_3) \tag{1.13}$$

$$c_G^{-1} = [(11/3)C_{ad} - 2C_f/3]/(16\pi)^3 \tag{1.14}$$

where C_{ad} is the dimension of the fundamental representation of the group and C_f is the number of fermions (fermion flavor) of the interaction.[12]

The known vector interactions and coupling constants of the Standard Theory are:[13]

- The Strong interaction coupling constant[14] $e_S = 1.22$
- The Weak SU(2) coupling constant $e_W = 0.619$
- The Electromagnetic U(1) coupling constant $e_{QED} = 0.303$

We found good approximations to the SU(2) and SU(3) coupling constants in Blaha (2019b).

---

[12] See Blaha (2019b) for the SU(2) and SU(3) values of c_G and the plots of all Standard Model eigenvalue functions.

[13] All coupling constant values are based on data extracted from C. Patrignani *et al* (Particle Data Group), Chinese Physics **C40**, 100001 (2014).

[14] Based on the running coupling constant value $\alpha_s(M_Z^2) = 0.1193 \pm 0.0016$.

The gauge interaction coupling constants, which we denoted with the label G with $e_G = (4\pi\alpha_G)^{1/2}$, for QED, Weak SU(2) and Strong SU(3) have a remarkable regularity—they double from interaction to interaction:[15]

The deeper significance of this regularity is not known.

Group	*Known Coupling Constant* e_G	*Known* $e_G^2/(4\pi)$	*Calculated* $\alpha_G = e_G^2/(4\pi)$	*Calculated[16] Exponent* g_G
QED, U(1)	0.30282212	$\alpha^{-1} = 137.035999084$	$\alpha^{-1} = 137.035999084$	−0. 00058053691948
SU(2)	0.619	0.0305	0.0425	0.54
SU(3)	1.22	0.118	0.086	0.5605

The relative closeness of the calculated values of "fine structure constants" to the experimentally known values is very encouraging—particularly in the case of the Electromagnetic fine structure constant α. It puts to rest other possible explanations for its value.

Our QED calculation of α has no free (adjustable) parameters unlike other attempts in the past. It also is totally based on Quantum Field Theory. The calculation of the non-abelian coupling constants also has no free (adjustable) parameters.

Thus the coupling constant eigenfunctions depend only on inherent perturbation theory based on dynamics. Coupling constant values cannot be "tweaked" to their known values by adjusting input parameters.

The ability of our 1973 calculation of the JBW eigenvalue function together with the new insights into understanding of the precise method to obtain its "fine structure constant" eigenvalues is also encouraging. It opens the possibility that the Standard Model has within itself the mechanism for determining the constants appearing within it. It raises the hope that a similar self-determination mechanism may

---

[15] Chapter 4 shows the universe scale factor g is ½ of the QED Fine Structure g deepening the mystery.
[16] They appear in eqs. 5.5 – 5.8 in Blaha (2019b). See Blaha (2019b) for more details.

also exist within the theory to determine the masses appearing in the Higgs particles sector of the theory.

3.1 Standard Model in All Universes

Since we will see in Chapter 4 that 4D universe(s) have the same evolutionary pattern of expansion and contraction, and since the Standard Model[17] coupling constants (which govern all macroscopic and chemical interactions) are determined internally within quantum field theory we can assert that the Physics, Chemistry, and Biology of all 4D universes are the same.

---

[17] One could suggest that the Standard Models of other universes are different. However the simplicity of the group structure would argue otherwise—as would consideration of the case of colliding universes.

4. Universe Coupling Constant and Dark Energy Vacuum Polarization

Perhaps the crowning achievement of our universal eigenvalue function formulation for coupling constants is the successful relation of universe evolution to vacuum polarization due to a vector QED-like interaction between universes.

4.1 Universal Scale Factor a(t)

The radiation-dominated and the matter-dominated scale factors a(t) both are power laws in time as seen earlier in Blaha (2019c). We therefore assumed[18] that the true a(t) has a power law form:

$$a(t) = (t/t_{now})^{g + ht} \tag{4.1}$$

where g and h are constants. (The constant h is *not* the Hubble parameter.) There is an "ht" term in the exponent based on the rise in H(t) suggested by experimental data.

The bases of this choice was:

1. Power law behavior (in part) as in the radiation and matter dominated approximations seen earlier.
2. The known shape of H(t) at early times, and at present, as seen and discussed in the Introduction. Eq. 22.1 is consistent with the shape of H(t).
3. The simplicity of the fit. Two values of H(t) set the constants g and h.
4. Faster than exponential future growth with no Big Rip.

$$a(t) = \exp[(g + ht)\ln(t/t_{now})] \sim e^{ht \ln(t)}$$

---

[18] Blaha (2019c).

The Hubble Constant implied by eq. 4.1 is

$$H(t) = (da/dt)/a = g/t + h(1 + \ln(t/t_{now})) \tag{4.2}$$

We set the value of $H(t)$ at two values of time determining g and h. Based on experimental data:

$$H(t_c) \equiv H(380,000 \text{ yr}) = 67.8 \tag{4.3}$$
$$H(t_{now}) = 73.24$$

and

$$h = (t_c H(t_c) - t_{now} H(t_{now}))[\, t_c - t_{now} + t_c \ln((t_c/t_{now})]^{-1} \tag{4.5}$$
$$g = (H(t_{now}) - h)\, t_{now}$$

where $t_c = 380,000$ years after the Big Bang we obtained

$$h = 2.25983 \times 10^{-18} \tag{4.6}$$

$$g = 0.000282377 = 2.82377 \times 10^{-4}$$

4.2 Massless QED

In massless QED we found that the vacuum polarization had the form:[19]

$$F_1(\alpha)(p/\Lambda)^{2g_{QED}} \tag{4.7}$$

where $F_1(\alpha)$ is the "eigenvalue function" for the Fine Structure Constant[20] of the Johnson-Baker-Willey model of massless QED, p is the momentum, and Λ is the ultraviolet cutoff. The value of g_{QED} that corresponded to the Fine Structure Constant is[21]

$$g_{QED} = -0.00058053691948 \tag{4.8}$$

---

[19] Eq. 12 in S. Blaha, Phys Rev **D9**, 2246 (1973).
[20] The author calculated $\alpha = 1/137...$ exactly (within experimental limits) in Blaha (2019a) and (2019b).
[21] Chapter 2 of this book.

and the Fine Structure Constant was correctly found (well within experimental limits) to be

$$\alpha_{calculated}(g_{QED}) = 0.0072973525693 \qquad (4.9)$$

to 13 digit accuracy according to the Particle Data Table of 2019.
Comparing our g value (eq. 4.6 above) with g_{QED} we

$$-g = 0.000282377 \cong -\tfrac{1}{2}g_{QED} = -0.000290268 \qquad (4.10)$$

4.3 Comparison of QED Vacuum Polarization Exponent with Universe Vacuum Polarization Exponent

Eq. 4.10 shows the numeric values of the g powers are approximately equal up to a factor of -2. The QED exponent describes high energy vacuum polarization behavior. The universe power g describes the small time universe expansion (near the Big Bang). The relation between the g powers, and the power function expressions of eq. 4.1 and 4.7, clearly show a close analogy.

Further the low energy (infrared) behavior of the QED vacuum polarization which is mass dependent is analogous to the large time (recent time) behavior of $a(t)$ which is governed by the h term in the exponent of $a(t)$.

The problem now before us is to find the universe vacuum polarization due to a new vector interaction between universes, and show that it is related to the QED vacuum polarization by eq. 4.10.[22]

4.4 A New Vector Interaction for Universe Particles

We assume universes can be treated as particles in 4-dimensional space-time.[23] Since experiments appear to have shown that our universe does not rotate (does not have spin)[24] we will assume the universe is a spin 0 boson. We assume that universes

---

[22] The following subsections appeared in Blaha (2019c).

[23] Universes are composite entities but we can treat them as quantum particles in the same manner as physicists treated protons and neutrons etc. as quantum particles before quark theory was accepted.

[24] The lack of universe rotation (spin) is indicated by a study of Cosmic Microwave Background (CMB) by D. Saadeh *et al*, Phys. Rev. Lett. **117**, 313302 (2016).

have a vector field interaction similar to QED. It is possible that the quantum vector $Y^\mu(x)$ field of the Big Bang quantum coordinates, treated previously, may be the vector field universe interaction field as well.

Given this QED-like framework, universe-antiuniverse pair production and vacuum polarization becomes possible. We assume the QED-like boson lagrangian

$$\mathcal{L} = \tfrac{1}{2}\,(\partial_\mu\varphi^\dagger\partial^\mu\varphi - m^2\varphi^\dagger\varphi) - ie_0\colon \varphi^\dagger(\overset{\rightarrow}{\partial_\mu} - \overset{\leftarrow}{\partial_\mu})\,\varphi\colon A^\mu + e_0^2\colon A^2\colon \colon\varphi^\dagger\varphi\colon + \delta m^2\colon\varphi^\dagger\varphi\colon$$

$$(4.11)$$

where $\varphi(x)$ is a "charged" quantum universe particle field.[25]

We now proceed to calculate the second order vacuum polarization of a universe particle.

4.5 Second Order Vacuum Polarization of a Universe Particle

The one loop vacuum polarization Feynman diagram is

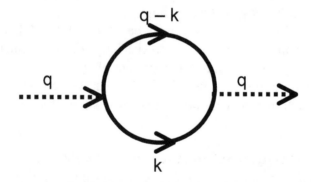

Figure 4.1 One loop vacuum polarization boson Feynman diagram.

Its evaluation is

---

[25] The charge is not electromagnetic charge.

$$I_{\mu\nu} = (-ie_0)^2 \int \frac{d^4k}{(2\pi)^4} \; \frac{i}{(k^2 - m^2 + i\varepsilon)} \; \frac{i}{(k^2 - m^2 + i\varepsilon)} \; (q - 2k)_\mu (q - 2k)_\nu \tag{4.12}$$

$$= \frac{\alpha}{2\pi} \int_0^\infty dz_1 \int_0^\infty dz_2 \; \frac{g_{\mu\nu} \exp[i(q^2 z_1 z_2/(z_1 + z_2) - (m^2 + i\varepsilon)(z_1 + z_2))]}{(z_1 + z_2)^3} + \text{gauge terms}$$

upon introducing parameters z_1 and z_2 to enable exponentiation and integration over k, where

$$\alpha = e_0^2/4\pi \tag{4.13}$$

Applying $q^2 \partial/\partial q^2$ to $I_{\mu\nu}$ to eliminate the quadratic divergent part, and then using the identity

$$1 = \int_0^\infty d\lambda/\lambda \; \delta(1 - (z_1 + z_2)/\lambda)$$

and letting $z_i = \lambda x_i$ we obtain

$$I_{\mu\nu} = \frac{i\,\alpha}{2\pi} \; q^2 g_{\mu\nu} \int dx_1 \int dx_2 \int d\lambda/\lambda \; x_1 x_2 \exp[i\lambda(q^2 x_1 x_2 - (m^2 + i\varepsilon))] \; \delta(1 - x_1 - x_2)$$

$$\tag{4.14}$$

up to gauge terms. The λ integration yields a logarithmic divergence which we cut off. Then

$$I_{\mu\nu} = \frac{i\,\alpha}{2\pi} \; q^2 g_{\mu\nu} \int_0^1 dx \; x(1\text{-}x) \ln(q^2 x(1 - x) - m^2) + \ldots \tag{4.15}$$

which becomes

$$I_{\mu\nu} = \frac{i\,\alpha}{12\pi} \; q^2 g_{\mu\nu} \ln(\Lambda^2/m^2) + \ldots \tag{4.16}$$

with finite and other gauge terms not shown.

Thus we find the renormalization constant Z_{3U} for the scalar universe particle case to be

$$Z_{3U} = 1 - \alpha/12\pi \ln(\Lambda^2/m^2) \qquad (4.17)$$

If we let

$$\alpha_U = \alpha/4 \qquad (4.18)$$

then we obtain the form similar to the one loop value of Z_3 for spin ½ electron QED:

$$Z_{3U} \cong 1 - \alpha_U/3\pi \ln(\Lambda^2/m^2) \qquad (4.19)$$

We now *provisionally assume* that α is the QED fine structure constant. We denote it as α_{QED}. We verify this choice later.

Thus the "fine structure constant" α_U for our vector interaction is

$$\alpha_U \equiv \alpha_{QED}/4 = 0.001824338 \qquad (4.20)$$

We now turn to the Johnson-Baker-Willey (JBW) model of massless QED since at ultra-high energy our vector interaction theory with lagrangian eq. 25.6 becomes the JBW model. In the JBW model we calculated α_{QED} and found the corresponding power which we denote g_{QED}.

Now we perform the same calculation for universe vacuum polarization and find the g value denoted g_U corresponding to α_U. The value of g_U will be seen to lead to the power g in the universal scale factor almost exactly.

The universe eigenvalue function is[26]

$$F_2(\alpha_U) = F_1(\alpha_U) - [2/3 + \alpha_U/(2\pi) - (1/4)[\alpha_U/(2\pi)]^2] \qquad (4.21)$$

For

$$\alpha_U \equiv \alpha_{QED}/4 = 0.001824338 \qquad (4.22)$$

we found the eigenfunction value

---

[26] We assume the universe eigenvalue function has the same form as the QED eigenvalue function.

$$F_2(\alpha_U = 0.001824338) = 5.10824 \times 10^{-12} \cong 0 \tag{4.23}$$

Examining $F_2(\alpha_U)$[27] as a function of g_U we found the value of g_U corresponding to α_U is

$$g_U = -0.00014525 \tag{4.24}$$

so the universe vacuum polarization is

$$\Gamma_U(p) = (p/\Lambda)^{2g_U} \tag{4.25}$$

The fourier transform is[28]

$$a(t) = (1/2\pi) \int_0^\infty dp/p \; \exp(-ipt) \; \Gamma_U(p) \tag{4.26}$$

$$= k \, (t/T)^{-2g_U} \tag{4.27}$$

where k is a constant and where

$$1/T = \Lambda \tag{4.28}$$

with Λ being the "momentum space" cutoff mass. Comparing eq. 4.1 and 4.27 we find

$$\begin{aligned} g &= -2g_U \\ &= 0.0002905 \end{aligned} \tag{4.29}$$

From eq. 4.6 for the power g of a(t) we see the universal scale factor g is

$$g = 0.000282377 \tag{4.30}$$

---

[27] $F_2(\alpha_U)$ and $F_2(g_U)$ are alternate notations for the same function.
[28] Those who might object to fourier transforming to time t should remember that inside a Black Hole the "time-like" coordinate is the radius and the time variable t is comparable to a spatial coordinate. The possibility that the universe is a Black Hole is not excluded. This fourier transform appears in Blaha (2019c) in eq. 25.25 with a typographic error—the division by p was omitted.

Thus the value of g calculated from the universe vacuum polarization differs from the actual value of g by less than 3%. Given the approximate nature of our JBW calculation of vacuum polarization the agreement is remarkable.[29]

In addition we found the "fine structure constant" for the vector interaction to be given by eq. 4.22 resulting in

$$e_U = (4\pi\alpha_U)^{\frac{1}{2}} = 0.151411 \tag{4.31}$$

Thus we have shown the universe vacuum polarization $\Gamma_U(p)$ when transformed to time is the universal scale factor a(t) up to a constant. The evolution of our universe is set by universe vacuum polarization. Other 4D universes may be expected to be similar.

The above relation we have found between QED-like vacuum polarization and universe vacuum polarization (Dark Energy) appears to clinch our interpretation of universe Dark Energy as, in the main, a consequence of universe vacuum polarization due to a universe vector interaction.[30]

4.6 Dark Energy is Equivalent to Universe Vacuum Polarization

Dark Energy is elusive both on the experimental and theoretical levels. We know it exists through its effects on our universe. Yet interactions with matter have not been found. Thus it is somewhat of a phantom.

The existence of Dark Energy, which, clearly, strongly affects the evolution of the universe, means that the Einstein equation, usually regarded as central to universe evolution, is incomplete for that purpose. It does not specify the total energy density ρ_{tot}.

---

[29] And may be exact! The value of the Hubble Constant H in recent times varies from about 70 – 75 making the calculation of g also approximate. We chose an average value of 73.24 to obtain the value of g above. If we chose the current value for H to be 75.58 we would have g = -2g_U exactly (eq. 25.28). Note: studies of binary black hole merger gravity waves have given a Hubble Constant of 75.2 km s^{-1} Mpc^{-1} (and earlier of 78 km s^{-1} Mpc^{-1}), and studies of light bent by distant galaxies give H = 72.5 km s^{-1} Mpc^{-1}. Thus the value H = 75.58 is not unreasonable. See section 22.1 for a summary of studies of H.

[30] Rather like the discovery of the Ω^- particle in the 1960s confirmed Gell-Mann's SU(3) theory.

$$\dot{a}^2 - 8\pi G\rho_{tot}a^2/3 = -k \qquad (4.32)$$

However we can obtain a "handle" on the total energy density by inserting our universal scale factor a(t) in the Einstein equation together with the known radiation density, matter density and Cosmological Constant Λ terms:

$$\rho_{tot}(t) \equiv \rho_{crit}\Omega_{tot}(t) = \rho_{crit}[\Omega_\Gamma(t) + \Omega_M(t) + \Omega_\Lambda + \Omega_T]$$

where the unknown part needed to makes the Einstein equation correct is the elusive Dark Energy $\rho_T(t)$

$$\rho_T(t) = \rho_{crit}\Omega_T(t) \qquad (4.33)$$

Then we can calculate energy density $\rho_{Dark}(t)$ as a function of time as well as related quantities as the following plots show.

4.7 Quasi-Free Universe Particles

Since $F_2 \cong 0$ by eq. 4.23 universe particles are very much like free particles since the vacuum polarization is zero except for a divergence due to the effect of the three subtracted terms displayed in eq. 4.21.

Universe particles are not totally free particles due to gravitation and Standard Model interactions such as electromagnetism. We treated the case of free universe particles in Blaha (2018).

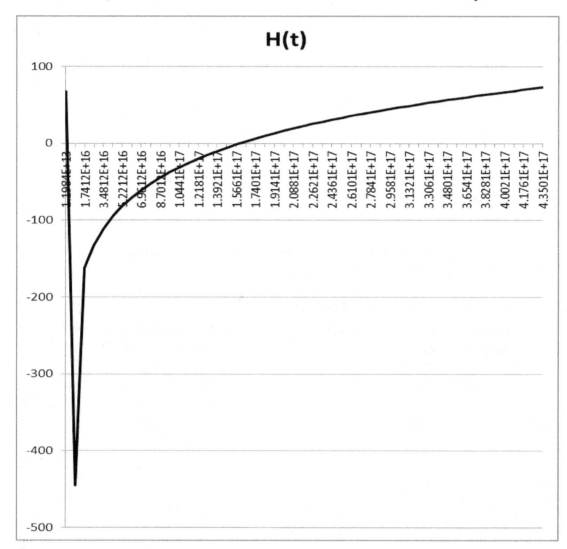

Figure 4.2. The Hubble Constant H(t) plotted vs. seconds from t = 1.19×10^{13} sec. to the present 4.35×10^{17} sec. The minimum is H = -445 km s^{-1} Mpc^{-1} at t = 8.71×10^{13} sec.

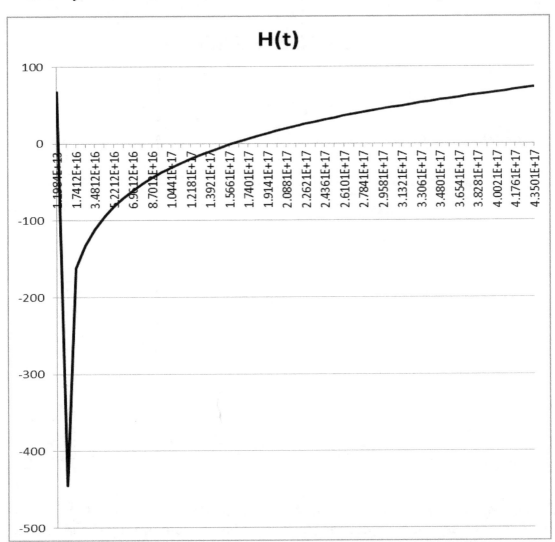

Figure 4.3. A closer view of H(t) plotted vs. seconds from t = 1.19 × 10^{13} sec. to the present 4.35 × 10^{17} sec. The minimum is H = -445 km s^{-1} Mpc^{-1} at t = 8.71 × 10^{13} sec.

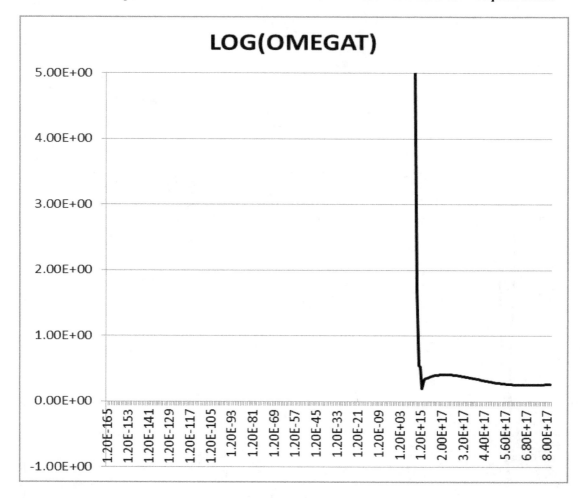

Figure 4.4. Log $\Omega_T(t)$ plotted vs. log time in seconds from 1.19×10^{-165} sec. to the 8.2×10^{17} sec. (almost double the present time). Note the Big Dip in $\Omega_T(t)$ at about t = 8.71×10^{13} sec. followed by a rise then a decline. At t = 1.19×10^{-165} sec. we found (not shown) log $\Omega_T(t) \approx 358$, an enormous value, similar to the level of vacuum energy found in quantum field theory. It declines to the plotted data shown above.

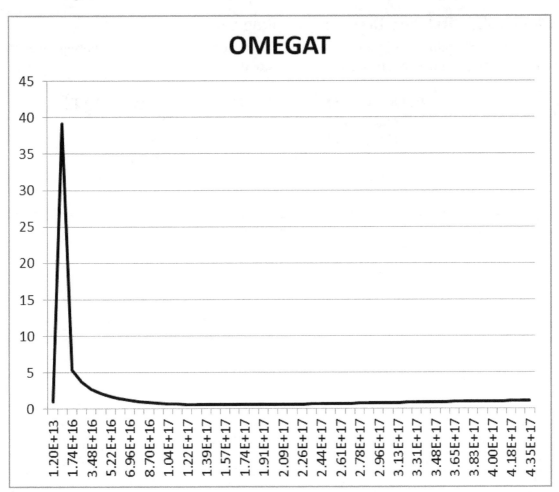

Figure 4.5. $\Omega_T(t)$ plotted vs. time in sec. from the year 380,000 to the present. Note the peak at $t = 8.71 \times 10^{13}$ sec. suggesting an influx into the universe at the transition to the matter-dominated phase, and then a decline followed by a raise. The peak value of $\Omega_T(t)$ is 39.2.

4.8 Doubling Relation Between Coupling Constants

The coupling constants that we have derived show a doubling whose fundamental significance remains to be understood.

INTERACTION	COUPLING CONSTANT[31]
Y Interaction e_U	0.1514
QED $e_{QED} = (4\pi\alpha_{QED})^{\frac{1}{2}}$	0.303
Weak SU(2) g_W	0.619
Strong SU(3) g_S	1. 22

Figure 4.6. The interaction coupling constants show a regular doubling. A fundamental cause for doubling is not apparent.

---

[31] M. Tanabashi *et al* (Particle Data Group), Phys. Rev. D**98**, 030001 (2018).

5. Major Implications For Universes

Our ability to calculate each of the Standard Model coupling constants and the universe vector interaction coupling constant (and universe scale factor) raises important issues both within our universe and in possible other universes. (The experimental and theoretical basis of the possibility of other universes is described in earlier books such as Blaha (2018).)

In our universe we see that the Anthropic hypothesis, based on the precise value of the QED Fine Structure Constant, is not compelling. The value of α is set within QED not otherwise. In addition the other Standard Model coupling constants, which were found to good approximation in our approach, are also *self-determined* in Quantum Field Theory. Since the Standard Model interactions are responsible for the vast majority of the features of Physics, Chemistry and Life we find the universe's detailed structure is set by Quantum Field Theory.

If, as we propose, there are other four-dimensional (4D) universes within a higher dimension space called the Megaverse, then, assuming that the overall physical theory of a 4D universe[32] satisfies

1. The universe is described by General Relativity.

2. The particle interactions in the universe are those of the Standard Model (possibly extended).

we find 4D universes have interactions with the same coupling constants found in our work. We further find interaction coupling constants are independent of space and time—consistent with experiment.

---

[32] And possibly of other universes of different dimension.

The conclusions that follow from this line of reasoning are:

1. The Standard Model features are the same in all 4D universes—particularly the features of electromagnetic phenomena.

2. The scale factor and evolution of other universes parallels that of our universe. We therefore expect Dark Energy dominance, varying Hubble Constants; Big Dips, Superclusters, and voids in other universes. (The total energy of universes may differ.)

3. Other universes have the same Physics, Chemistry and Biology as our universe. Life in other universes is possible but can be expected to take different superficial forms.

4. Other universes can differ in some ways that do not change our overall conclusions: in some universes a) anti-particles may dominate; b) left-right symmetry may differ; c) Life may favor either dextrorotary or levorotary molecules;

5. Communication with creatures in other universes may be possible using more advanced quantum entanglement. Quantum entanglement of photons from earth and the sun has been found experimentally.

6. Universes have a composite particle nature and can be described in a second quantized framework. See Blaha (2018), and comments there by DeWitt in particular on quantum universe features.

REFERENCES

Blaha, S., 2018, *Unification of God Theory and Unified SuperStandard Model THIRD EDITION* (Pingree Hill Publishing, Auburn, NH, 2018).

_____, 2019a, *Calculation of: QED α = 1/137, and Other Coupling Constants of the Unified SuperStandard Theory* (Pingree Hill Publishing, Auburn, NH, 2019).

_____, 2019b, *Coupling Constants of the Unified SuperStandard Theory SECOND EDITION* (Pingree Hill Publishing, Auburn, NH, 2019).

_____, 2019c, *Quantum Big Bang-Quantum Vacuum Universes (Particles)* (Pingree Hill Publishing, Auburn, NH, 2019).

Gradshteyn, I. S. and Ryzhik, I. M., 1965, *Table of Integrals, Series, and Products* (Academic Press, New York, 1965).

Heitler, W., 1954, *The Quantum Theory of Radiation* (Claendon Press, Oxford, UK, 1954).

Huang, Kerson, 1992, *Quarks, Leptons & Gauge Fields 2nd Edition* (World Scientific Publishing Company, Singapore, 1992).

INDEX

CPSIA information can be obtained
at www.ICGtesting.com
Printed in the USA
BVHW050102101019
560654BV00010B/386/P